ARBEITSGEMEINSCHAFT FÜR FORSCHUNG
DES LANDES NORDRHEIN-WESTFALEN

NATUR-, INGENIEUR- UND GESELLSCHAFTSWISSENSCHAFTEN

180. SITZUNG
AM 4. JUNI 1969
IN DÜSSELDORF

ARBEITSGEMEINSCHAFT FÜR FORSCHUNG
DES LANDES NORDRHEIN-WESTFALEN

NATUR-, INGENIEUR- UND GESELLSCHAFTSWISSENSCHAFTEN

HEFT 200

MICHAEL F. ATIYAH

Vector Fields on Manifolds

HERAUSGEGEBEN
IM AUFTRAGE DES MINISTERPRÄSIDENTEN HEINZ KÜHN
VON STAATSSEKRETÄR PROFESSOR Dr. h. c. Dr. E. h. LEO BRANDT

MICHAEL F. ATIYAH

Vector Fields on Manifolds

Springer Fachmedien Wiesbaden GmbH

ISBN 978-3-322-97941-4   ISBN 978-3-322-98503-3 (eBook)
DOI 10.1007/978-3-322-98503-3

© 1970 by Springer Fachmedien Wiesbaden

Ursprünglich erschienen bei Westdeutscher Verlag, Köln und Opladen 1970

Gesamtherstellung: Westdeutscher Verlag GmbH ·

Inhalt

Michael F. Atiyah, Ph. D., F.R.S., Savilian Professor of Geometry, Oxford University*

Vector Fields on Manifolds

§ 1 Introduction 7
§ 2 Clifford algebras and differential forms 8
§ 3 Euler characteristic and signature 11
§ 4 Kervaire semi-characteristic 15
§ 5 Vector fields with finite singularities 18

References 24

Zusammenfassung 25

Résumé 26

---

* Now at the Institute for Advanced Study, Princeton.

§ 1 Introduction

This paper is a contribution to the topological study of vector fields on manifolds. In particular we shall be concerned with the problems of existence of r linearly independent vector fields. For $r = 1$ the classical result of H. Hopf asserts that the vanishing of the Euler characteristic is the necessary and sufficient condition, and our results will give partial extensions of Hopf's theorem to the case $r > 1$. A recent article by E. Thomas [10] gives a good survey of work in this general area.

Our approach to these problems is based on the index theory of elliptic differential operators and is therefore rather different from the standard topological approach. Briefly speaking, what we do is to observe that certain invariants of a manifold (Euler characteristic, signature, etc.) are indices of elliptic operators (see [5]) and the existence of a certain number of vector fields implies certain symmetry conditions for these operators and hence corresponding results for their indices. In this way we obtain certain necessary conditions for the existence of vector fields and, more generally, for the existence of fields of tangent planes. For example, one of our results is the following

THEOREM (1.1). *Let X be a compact oriented smooth manifold of dimension $4q$, and assume that X possesses a tangent field of oriented 2-planes (that is, an oriented 2-dimensional sub-bundle of the tangent vector bundle). Then the Euler characteristic of X is even and is congruent to the signature of X modulo 4.*

Of course, as a corollary of (1.1), we deduce that the existence of 2 independent vector fields implies that the signature of X is divisible by 4. Generalizing this, we show that the existence of r independent vector fields implies divisibility of the signature by an appropriate power of 2 (depending on r).

In addition to using the index of an elliptic operator we also use the mod 2 index of a real skew-adjoint elliptic operator (see [6]). The real Kervaire semi-characteristic of a $4q + 1$-manifold, defined by

$$k(X) = \sum_p \dim_R H^{2p}(X; R) \mod 2$$

is an example of such a mod 2 index, and using this we shall prove

THEOREM (1.2). *Let X be a compact oriented smooth manifold of dimension $4q+1$, and assume that X possesses a tangent field of oriented 2-planes. Then $k(X) = 0$.*

On odd-dimensional oriented manifolds the Euler characteristic vanishes and so one vector field v_1 (without zeros) always exists. Theorem (1.2) shows that $k(X)$ is an obstruction to the existence of a second independent vector field v_2. If v_2 exists then (by rotation in the (v_1, v_2)-plane) we see that v_1 is homotopic to $-v_1$. If $k(X) \neq 0$ then v_2 does not exist and in fact we can prove

THEOREM (1.3). *Let X be compact oriented and of dimension $4q+1$ and assume $k(X) \neq 0$. Then a (nowhere zero) vector field v on X is never homotopic to $-v$.*

Theorems (1.2) and (1.3) apply for example to the sphere S^{4q+1}, since $k(S^{4q+1}) = 1 \neq 0$.

To understand the topological significance of our analytical methods one must pass from operators to symbols and then employ K-theory as developed in [4]. Once expressed in the framework of K-theory our methods can moreover be refined to deal with r fields with finite singularities in the sense of [10]. These refinements, which will be briefly explained in § 5, will be expounded in greater detail elsewhere.

All our analysis is best expressed in the context of Clifford algebras and in § 2 we review the relevant material in this direction. In § 3 we study the Euler characteristic and signature and, in particular, we prove Theorem (1.1). In § 4 we deal with the Kervaire semi-characteristic and prove Theorems (1.2) and (1.3).

The results presented here are part of the general investigations into the index theory of elliptic operators being undertaken jointly with I. M. Singer.

§ 2 Clifford algebras and differential forms

For a real vector space V of dimension n we have the exterior algebra

$$\wedge^*(V) = \sum_{p=0}^{n} \wedge^p(V)$$

If V is Euclidean (i.e. if V is endowed with a positive definite quadratic form Q) we can form the Clifford algebra $C(V; -Q)$ (see [2]). This algebra contains V and for any $v \in V$ we have

$$v^2 = -Q(v) 1,$$

this being the defining identity of the Clifford algebra. When Q is understood we write simple $\text{Cliff}(V)$ instead of $C(V; -Q)$.

As vector spaces both $\wedge^*(V)$ and $\text{Cliff}(V)$ have dimension 2^n. In fact there is a natural vector space isomorphism

$$\varphi: \text{Cliff}(V) \to \wedge^*(V)$$

obtained in the following way. For each $v \in V$ let $A(v): \wedge^*(V) \mapsto \wedge^*(V)$ be the linear transformation defined by

$$A(v) w = v \wedge w.$$

Now, using the Euclidean inner product on V, we obtain a natural inner product on $\wedge^*(V)$ and hence we can define the adjoint transformation $A(v)^*$ of $A(v)$. If we put

$$L(v) = A(v) - A(v)^*$$

it is easy to verify that

$$L(v)^2 = -Q(v) \cdot I$$

where I is the identity automorphism of $\wedge^*(V)$. Hence $v \mapsto L(v)$ extends to an algebra homomorphism

$$L: \text{Cliff}(V) \to \text{End}(\wedge^*(V))$$

where $\text{End}(\wedge^*(V))$ denotes the $2^n \times 2^n$ matrix algebra of linear endomorphisms of the vector space $\wedge^*(V)$. Finally we define φ by $\varphi(x) = L(x)1$, where $1 \in \wedge^0(V)$ is the identity of the exterior algebra, and $x \in \text{Cliff}(V)$. In terms of an orthonormal base $(e_1, \ldots, e_n)$ of V we have

$$\varphi(e_{i_1} \ldots e_{i_p}) = e_{i_1} \wedge \ldots \wedge e_{i_p} \qquad (i_1 < i_2 < \ldots < i_p)$$

and so φ is an isomorphism. Note that, for $x \in \text{Cliff}(V)$ and $v \in V$ we have

$$L(v) \varphi(x) = \varphi(vx)$$

so that, if we identify $\text{Cliff}(V)$ and $\wedge^*(V)$ by means of φ, $L(v)$ becomes *left Clifford multiplication* by v.

Suppose now that X is a Riemannian manifold, then its tangent spaces $T_x (x \in X)$ are Euclidean and can be identified with their duals T_x^*. The exterior algebras $\wedge^*(T_x^*)$ for $x \in X$ form a bundle of algebras over X, and so also do the Clifford algebras $\text{Cliff}(T_x^*)$. The natural vector space isomorphisms

$$\varphi_x: \text{Cliff}(T_x^*) \to \wedge^*(T_x^*)$$

induce a vector bundle isomorphism

$$\text{Cliff}(T^*X) \cong \wedge^*(T^*X).$$

Identifying these two vector bundles via this isomorphism we may say that the bundle $\wedge^*(T^*X)$ has two different algebra structures – the exterior algebra and Clifford algebra. Thus if α, β are two exterior differential forms on X we can define their *exterior* product $\alpha \wedge \beta$ and also their *Clifford* product $\alpha \cdot \beta$. Both products are associative. Note that $\alpha \wedge \beta$ is independent of the Riemannian metric while $\alpha \cdot \beta$ depends on the metric. Also if α, β are homogeneous, i.e. $\alpha \in \wedge^p$, $\beta \in \wedge^q$, then $\alpha \wedge \beta$ is also homogeneous ($\alpha \wedge \beta \in \wedge^{p+q}$) but $\alpha \cdot \beta$ is not homogeneous: in fact it is a sum of terms of degrees $p + q - 2k$ for $k = 0, 1, \ldots$

The basic operator on differential forms for our purposes is $d + d^*$, where d is the exterior derivative and d^* is its formal adjoint. Since

$$(d + d^*)^2 = dd^* + d^*d = \Delta$$

is the Laplace operator it follows that $d + d^*$ is elliptic and it is clearly formally self-adjoint. By a standard argument [5; § 6] this implies that, for compact X,

$$\text{Ker}(d + d^*) = \text{Ker}\,\Delta$$

consists of the harmonic forms, which by the Hodge theory can be identified with the real cohomology of X.

The symbol of d is i times exterior multiplication, that is $\sigma_d(\xi) = iA(\xi)$ for $\xi \in T_x^*$. Hence, taking adjoints, we get $\sigma_{d^*}(\xi) = -iA(\xi)^*$ and so

(2.1) $\quad \sigma_{d+d^*}(\xi) = i\{A(\xi) - A(\xi)^*\} = iL(\xi)$

where $L(\xi)$ is, as above, left Clifford multiplication by ξ.

Suppose now that u is a vector field on X which we can also regard as a 1-form by using the Riemannian metric. Then we can consider the operation $R(u)$ on differential forms given by *right* Clifford multiplication by u. Since $R(u)^2 = R(u^2) = -|u|^2 I$ it follows that, if u is nowhere zero, $R(u)$ is an automorphism of the space of differential forms. Now Clifford multiplication of forms being associative, it follows that left and right multiplications commute and so, by (2.1), $R(u)$ commutes with σ_{d+d^*}. This means that

(2.2) $\quad (d+d^*) \circ R(u) - R(u) \circ (d+d^*)$ *is of order zero,*

the first derivatives all cancelling. This fact is of cardinal importance for all our applications and is the way in which vector fields interact with the analysis.

In the next two sections we shall pursue this matter further, treating separately various values of n mod 4.

§ 3 Euler characteristic and signature

As a simple illustration of our methods we shall first of all give an analytical proof of part of Hopf's theorem, namely that the existence of a nowhere zero vector field implies the vanishing of the Euler characteristic*.

First we recall that the Euler characteristic $E(X)$ is the index of the elliptic operator

$$D: \Omega^{\text{ev}} \to \Omega^{\text{odd}}$$

obtained by restricting $d + d^*$ to even forms. This is clear (by the Hodge theory) because D^* is the restriction of $d + d^*$ to odd forms and so

$$\text{index } D = \dim H^{\text{ev}} - \dim H^{\text{odd}}$$

where H is the space of harmonic forms.

Now we use our nowhere zero vector field u to define an automorphism $R(u)$ of the space of forms as in § 2. This automorphism interchanges Ω^{ev} and Ω^{odd} and in virtue of the approximate commutation (2.2) we see that

$$R(u)^{-1} D R(u) - D^*$$

is of order zero. Hence

$$\text{index } D = \text{index } R(u)^{-1} D R(u)$$
$$= \text{index } D^* \text{ (since 0-order terms do not alter the index)}$$
$$= -\text{index } D$$

and so $E(X) = \text{index } D = 0$ as required.

* The argument which follows was briefly indicated in [1].

We now pass on to consider the signature S of an oriented $4q$-dimensional manifold. As explained in [5; § 6] this is also the index of a certain elliptic operator D^+. We recall briefly the definition of D^+. We define an involution τ on forms by

$$\tau(\alpha) = (-1)^{q+p(p-1)/2} * a \qquad (\alpha \in \Omega^p).$$

Then τ anti-commutes with $d + d^*$ and so the restriction of $d + d^*$ defines an elliptic operator

$$D^+ : \Omega^+ \to \Omega^-$$

where $\Omega^\pm$ are the ± 1-eigenspaces of τ.

If we put

$$\Omega^\pm_{ev} = \Omega^{ev} \cap \Omega^\pm, \quad \Omega^\pm_{odd} = \Omega^{odd} \cap \Omega^\pm$$

then $d + d^*$ restricts to give two elliptic operators

$$A_+ : \Omega^+_{ev} \to \Omega^-_{odd}$$
$$A_- : \Omega^-_{ev} \to \Omega^+_{odd}.$$

Clearly we have

$$\text{index } A_+ + \text{index } A_- = \text{index } D = E$$
$$\text{index } A_+ - \text{index } A_- = \text{index } D^+ = S.$$

Hence

$$\text{index } A_+ = \tfrac{1}{2}(E + S)$$
$$\text{index } A_- = \tfrac{1}{2}(E - S)$$

For our applications these two operators are rather more refined than the operators D^+, D and give better results.

The *-operator and the involution τ can be expressed in terms of Clifford multiplication as follows. Let w be the $4q$-form on X defined by the orientation and the metric ($w = *1$), and let $L(w)$ denote as before left Clifford multiplication by w. Then for a p-form α we have

$$L(w)\alpha = (-1)^{p(p-1)/2} * \alpha$$
$$= (-1)^q \tau \alpha$$

Thus $L(w) = (-1)^q \tau$ so that $\Omega^\pm$ are also the eigenspaces of $L(w)$.

We are now ready to give *the proof of Theorem* (1.1), so let u be the 2-form given by the field of oriented 2-planes on X, and $R(u)$ the operation of right Clifford multiplication by u. Then $R(u)$ preserves the spaces $\Omega^\pm$

(because $R(u)$ commutes with $L(w)$) and it also preserves parity of forms. Hence it preserves $\Omega_{ev}^{\pm}$ and $\Omega_{odd}^{\pm}$. In terms of a local orthonormal basis u_1, u_2 of the 2-plane we have $u = u_1 \cdot u_2$ and so $R(u) = R(u_1) R(u_2)$. Hence $R(u)^2 = -1$ and, by (2.2), $R(u)$ commutes with the operators $A_{\pm}$ modulo 0-order terms. Hence

$$B_{\pm} = \tfrac{1}{2} \{A_{\pm} + R(u) A_{\pm} R(u)^{-1}\}$$

commutes with $R(u)$ and has the same symbol as $A_{\pm}$ and therefore also the same index. On the other hand, the real vector spaces $\operatorname{Ker} B_{\pm}$, $\operatorname{Ker} B_{\pm}^*$ now admit the transformation $R(u)$ of square -1 and hence have even dimensions. Thus

$$\tfrac{1}{2}(E \pm S) = \operatorname{index} A_{\pm} = \operatorname{index} B_{\pm}$$
$$= \dim \operatorname{Ker} B_{\pm} - \dim \operatorname{Ker} B_{\pm}^*$$
$$\equiv 0 \bmod 2.$$

From these two congruences it follows that E is divisible by 2 and that $E \equiv S \bmod 4$. This completes the proof of Theorem (1.1).

It is clear that the method of proof of Theorem (1.1) yields at once a number of further generalizations. Thus the oriented 2-plane could be replaced by an oriented p-plane. The only requirements on the element u defined by the p-plane are that it be even and that $R(u)^2 = -1$: these amount to the condition $p \equiv 2 \bmod 4$. Thus as a generalization of (1.1) we have

THEOREM (3.1). *If the compact oriented 4 q-manifold X possesses a tangent field of oriented p-planes, with $p \equiv 2 \bmod 4$, then the Euler characteristic of X is even and is congruent modulo 4 to the signature of X.*

If X possesses 2 linearly independent vector fields then $E(X) = 0$ and, from Theorem (1.1), we deduce that the signature of X is divisible by 4. This result can be generalized further on the following lines. Assume that $v_1, \ldots, v_r$ are r linearly independent vector fields (which we may take to be orthonormal). Then we obtain operators $R(v_1), \ldots, R(v_r)$ on the space of forms which satisfy the Clifford identities

(3.2) $\quad R(v_i)^2 = -1, \quad R(v_i) R(v_j) = -R(v_j) R(v_i) \quad \text{for } i \neq j.$

Since $R(v_i)$ commutes with $L(w)$ it preserves the spaces $\Omega^{\pm}$. Since, by (2.2), $R(v_i)$ commutes with D^+ modulo 0-order terms, it follows that the elliptic operator T, formed by averaging D^+ over the finite group generated by the $R(v_i)$, has the same symbol and index as D^+. Moreover T now commutes strictly with the $R(v_i)$ and hence Ker T and Ker T^* become Cliff(R^r)-modules. Grading Ker T and Ker T^* by the parity of forms, we see that these are both Z_2-*graded* Clifford-modules in the sense of [2], and hence their dimensions are divisible by $2 a_r$ where a_r is given by [2; Table 2]. Thus finally we have

$$S = \text{index } D^+ = \text{index } T = \dim \text{Ker } T - \dim \text{Ker } T^* \equiv 0 \bmod 2\, a_r.$$

In the preceding argument the vector fields v_i could equally well be replaced by fields of oriented p_i-planes where $p_i \equiv 1 \bmod 4$. The operators $R(v_i)$ still satisfy the Clifford identities (3.2). Thus we have proved

THEOREM (3.3). *Let X be a compact oriented $4q$-manifold and assume it possesses r independent fields of oriented planes $v_1, \ldots, v_r$, where $\dim v_i \equiv 1 \bmod 4$. Then the signature of X is divisible by $2\, a_r$ where the numbers a_r are given by*

$r =$	1	2	3	4	5	6	7	8
$a_r =$	1	2	4	4	8	8	8	8

and $a_{r+8} = 16\, a_r$.

Remarks. For the case of r vector fields similar results have been obtained by K. Mayer [9] and D. Frank [7]. In fact, Mayer's results* are the best in this direction and his methods are very close to those we have been using. Mayer uses the general index formula of [5] whereas we have been more elementary in our proof and simply used general properties of elliptic operators. Our divisibility results come from local symmetry properties of the symbol of an operator. It is rather interesting that, in certain dimensions, Mayer's results improve ours by a factor of 2. This extra factor has then a global origin and cannot be deduced from local considerations. We shall have more to say on this topic when we investigate r-fields with finite singularities. The methods of Frank are quite different and his results coincide with those obtained by Mayer's method, but only using complex K-theory: Mayer gets further refinements from real K-theory.

---

* Mayer's Theorem can be specialized down to give results on the signature. This is not explicitly carried out in Mayer's paper but has been done by Wilhelm Schwarz (Diplomarbeit, Bonn 1965).

§4 Kervaire semi-characteristic

For a compact oriented manifold X of *odd* dimension we define the (real) Kervaire semi-characteristic $k(X)$ by

$$k(X) = \sum_p \dim_R H^{2p}(X; R) \mod 2.$$

As we shall now show this has an analytical interpretation when $\dim X \equiv 1 \mod 4$.

Choosing a Riemannian metric we introduce, as in §3, the top-dimensional form w and the operation $L(w)$. Since $\dim X$ is odd we now find that

$$L(w)\varphi = (-1)^p * \varphi \qquad \text{if } \varphi \in \Omega^{2p}$$
$$L(w)\varphi = (-1)^{p+1} * \varphi \qquad \text{if } \varphi \in \Omega^{2p+1}$$

and $*^2 = 1$.

Moreover the adjoint d^* of d is given by

$$d^*\varphi = (-1)^p * d * \varphi \qquad \text{if } \varphi \in \Omega^p.$$

From these we deduce that, for $\varphi \in \Omega^{2p}$,

$$L(w)(d + d^*)\varphi = L(w)(d\varphi + * d * \varphi) = (-1)^{p+1} * d\varphi + (-1)^p d * \varphi$$
$$(d + d^*)L(w)\varphi = (-1)^p (d + d^*) * \varphi = (-1)^p d * \varphi + (-1)^{p+1} * d\varphi.$$

Thus $L(w)$ commutes with $d + d^*$ (on Ω^{2p}). Since $\dim X \equiv 1 \mod 4$ we have $L(w)^2 = -1$ and so (since $L(w)$ is unitary) $L(w)$ is skewadjoint. Since $d + d^*$ is self-adjoint it then follows that the operator

$$T: \sum \Omega^{2p} \to \sum \Omega^{2p},$$

given by $T(\varphi) = L(w)(d + d^*)\varphi$, is skew-adjoint. On the other hand, T, like $d + d^*$, is elliptic and

$$\operatorname{Ker} T = \sum H^{2p}$$

where H^{2p} is the space of harmonic $2p$-forms. Thus

$$k(X) = \dim \operatorname{Ker} T \mod 2$$

is the dimension modulo 2 of a real elliptic skew-adjoint operator. As explained in [6] this "mod 2 index" has stability properties like the ordinary index. Actually in [6] only bounded operators are considered whereas here we have to deal with differential operators, but we can reduce to the bounded case in the usual way as follows. Let $\wedge$ be the positive square root of

$1 + \Delta$ where $\Delta = (d + d^*)^2$ is the Laplacian (acting on $\sum_p \Omega^{2p}$). Then $\wedge$ has a compact inverse $\wedge^{-1}$ and $(d + d^*)\wedge^{-1}$ is bounded and self-adjoint. Moreover since $L(w)$ commutes with $d + d^*$ it commutes with $1 + \Delta$ and hence also with $\wedge$ and $\wedge^{-1}$. Thus $T\wedge^{-1}$ is bounded skew-adjoint and Fredholm.

We are now ready to give the *proof of Theorem* (1.2), so let u be the 2-form defined by the field of oriented 2-planes (and the metric), and let $R(u)$ be the operation of right Clifford multiplication by u. Since $R(u)$ has even degree, commutes with $L(w)$ and commutes modulo 0-order terms with $d + d^*$ (by (2.2)), it follows that $R(u)$ commutes with T modulo 0-order terms. Hence the operator

$$S = \tfrac{1}{2}\{T + R(u)TR(u)^{-1}\}$$

commutes with $R(u)$ and differs from T only in 0-order terms. Since $R(u)$ is unitary S is also skew-adjoint. Hence by the stability of the mod 2 index we see that

$$k(X) \equiv \dim \operatorname{Ker} T \equiv \dim \operatorname{Ker} S \quad \bmod 2.$$

But $\operatorname{Ker} S$ admits the transformation $R(u)$ and $R(u)^2 = -1$. Hence dim Ker S is even and so $k(X) = 0$ as required.

Remark. E. Thomas has obtained a result similar to (1.2) but using the mod 2 Kervaire semi-characteristic defined by

$$k_2(X) = \sum_p \dim_{Z_2} H^{2p}(X; Z_2).$$

He shows that if X is a spin-manifold of dimension $\equiv 1 \bmod 4$ and possesses 2 independent vector fields then $k_2(X) = 0$. Our Theorem (1.2) implies, with the same hypotheses, that $k(X) = 0$. The connection between these results is explained by an elegant formula of Lusztig–Milnor–Peterson [8] which, for any compact oriented $(4q + 1)$-manifold, asserts that

$$k(X) - k_2(X) = w_2 \cdot w_{4q-1}$$

where w_i is the i-th Stiefel–Whitney class of X. In particular, for a spin-manifold $w_2 = 0$ and so $k = k_2$. For manifolds with $w_2 \neq 0$ we have $k \neq k_2$. For such manifolds it is k, rather than k_2, which enters as an obstruction to finding 2 independent vector fields. From the point of view of conventional algebraic topology, it is perhaps a little surprising that, in detecting a mod 2 homotopy element (the obstruction), we should need real cohomology instead of mod 2 cohomology. However from the analytical standpoint the real numbers are obligatory! This seems to support the view that the analysis here is rather naturally involved with the problem of vector fields.

We pass next to the *proof of Theorem* (1.3). Let v be a nowhere zero vector field on X (which we may take of unit length). Then $R(v \cdot w)$ has even degree, commutes with T modulo 0-order terms and satisfies $R(v \cdot w)^2 = 1$. Averaging T we therefore produce a new skew-adjoint operator

$$Q = \tfrac{1}{2}\{T + R(v \cdot w)TR(v \cdot w)^{-1}\}$$

commuting with $R(v \cdot w)$ and such that

$$\dim \operatorname{Ker} Q \equiv \dim \operatorname{Ker} T \equiv k(X) \mod 2.$$

Decomposing $\operatorname{Ker} Q$ into the (± 1)-eigenspaces of $R(v \cdot w)$ we get two spaces of dimensions (mod 2) equal say to $a(v)$, $b(v)$. *These are homotopy invariants of v* (and independent of the choice of Riemannian metric). This follows from the stability of the mod 2 index under deformation. Moreover since $R(-v \cdot w) = -R(v \cdot w)$ it follows that

$$b(v) = a(-v)$$

and so

$$k(X) = a(v) + b(v) = a(v) + a(-v).$$

Assume now that $-v$ is homotopic to v, then $a(-v) = a(v)$ and so $k(X) \equiv 0 \mod 2$. This proves Theorem (1.3).

Just as in § 3 the methods of this section work also for fields of p-planes. Thus we get

THEOREM (4.1). *Assume that the compact oriented $(4q+1)$-manifold X admits a field of oriented p-planes with $p \equiv 2 \mod 4$. Then $k(X) = 0$.*

THEOREM (4.2). *Let v be a field of oriented p-planes, on the compact oriented $(4q+1)$-manifold X, with $p \equiv 1 \mod 4$. Assume that $k(X) \neq 0$. Then v is not homotopic to $-v$ (the same field but with the opposite orientation).*

Remark. Since $2 + 3 \equiv 1 \mod 4$, Theorem (4.1) also holds for $p \equiv 3 \mod 4$ (the orthogonal field then has dimension $\equiv 2 \mod 4$). Similarly (4.2) holds also for $p \equiv 0 \mod 4$.

§ 5 Vector fields with finite singularities

In this section I shall indicate briefly how the results of the preceding sections can be refined to give information about vector fields with finite singularities in the sense of Thomas [10]. This is done in terms of K-theory and requires the index theorem for elliptic operators in various forms.

Let us first recall the situation studied by Thomas. We consider r vector fields $v_1, \ldots, v_r$ on the closed oriented n-manifold X and we assume that they are linearly independent except at a finite set of points (the singularities). Then at each singular point A, by restricting $v_1, \ldots, v_r$ to a small ball around A, we obtain an element $a_A(v_1, \ldots, v_r) \in \pi_{n-1}(V_{n,r})$ where $V_{n,r}$ is the Stiefel manifold $SO(n)/SO(n-r)$. The element a_A represents the local obstruction to eliminating the singularity at A.

The general problem discussed by Thomas in [10] is to find global expressions for the sum $\sum_A a_A(v_1, \ldots, v_r)$ where A runs over all the singular points. When $r = 1$ we have $\pi_{n-1}(V_{n,n-1}) = \pi_{n-1}(S^{n-1}) \cong Z$ and a_A is the local degree. Hopf's theorem asserts that $\sum a_A$ is the Euler characteristic. For $r = 2$ we have*

$$\pi_{n-1}(V_{n,2}) = Z_2 \quad \text{for } n \text{ odd}$$
$$= Z \oplus Z_2 \quad \text{for } n \text{ even}$$

and for this case we can prove the following theorem:

THEOREM (5.1). *Let v_1, v_2 be 2 vector fields with finite singularities, then*

$$\sum_A a_A(v_1, v_2) = k(X) \quad \text{if } \dim X \equiv 1 \bmod 4$$
$$= 0 \quad \text{if } \dim X \equiv 3 \bmod 4$$
$$= E(X) \oplus 0 \text{ if } \dim X \equiv 2 \bmod 4$$
$$= E(X) \oplus \frac{E(X) - (-1)^k S(X)}{2} \text{ if } \dim X = 4k$$

Similar results have been obtained by quite different methods by E. Thomas and D. Frank (see [10] for more details).

Our methods also give some results for the case of general r but these will be explained elsewhere**. Here, as an indication of method, we shall

---
* We exclude from now on certain exceptional low values of n. We take $n > 4$.
** See also formula (5.4).

only prove the last part of (5.1). Moreover to simplify the exposition we shall restrict ourselves even further and deal only with a Spin-manifold of dimension divisible by 8: this avoids some of the technical complications, but involves all the main ideas.

We shall start by constructing a basic universal element

$$\mu_r \in KO(B\,\text{Spin}\,(8\,q) \times P_{r-1}, B\,\text{Spin}\,(8\,q-r) \times P_{r-1})$$

where $P_{r-1} = P(R^r)$ is $(r-1)$-dimensional real projective space and $B\,\text{Spin}\,(n)$ is the classifying space of $\text{Spin}\,(n)$. To do this we consider first the total Spin representation $\varDelta = \varDelta^+ \oplus \varDelta^-$ of $\text{Spin}\,(8\,q)$. This is the unique irreducible representation of the simple Clifford algebra $C_{8q} = \text{Cliff}\,(R^{8q})$. Since (see [2])

$$C_{8q} \cong C_{8q-r} \hat{\otimes} C_r,$$

where $\hat{\otimes}$ denotes the graded tensor product, it follows that C_r commutes with $\text{Spin}\,(8\,q-r) \subset C^0_{8q-r}$. Thus if $\varDelta$ is considered as a representation of $\text{Spin}\,(8\,q-r)$ its commuting algebra contains (and is actually equal to) C_r. Thus $\varDelta$ becomes a graded C_r-module commuting with the action of $\text{Spin}\,(8\,q-r)$. Passing to the associated bundles over the classifying spaces we see that $\varDelta$ defines a Z_2-graded bundle $M = M^+ \oplus M^-$ over $B\,\text{Spin}\,(8\,q)$ whose restriction to $B\,\text{Spin}\,(8\,q-r)$ is a bundle of Z_2-graded C_r-modules. From this data we produce the element α_r by a standard construction given essentially in [2; § 15]: it goes as follows. Writing B_n for $B\,\text{Spin}\,(n)$ we let $\pi: B_{8q} \times S^{r-1} \to B_{8q}$ be the projection and we consider the two bundles $\pi^* M^+$ and $\pi^* M^-$. We let the group Z_2 act on $B_{8q} \times S^{r-1}$ by the antipodal action on S^{r-1} and we cover this action by the trivial action on $\pi^* M^+$ and the non-trivial (i.e. -1) action on $\pi^* M^-$. Passing to the quotient $B_{8q} \times P_{r-1}$ this gives the bundles $M^+ \otimes 1$ and $M^- \otimes H$, where H is the Hopf bundle on P_{r-1}. Now over the subspace $B_{8q-r} \subset B_{8q}$ we have a graded C_r-structure on M. This Clifford multiplication gives an explicit isomorphism of $\pi^* M^+$ with $\pi^* M^-$ over $B_{8q-r} \times S^{r-1}$. Moreover, since Clifford multiplication is bilinear, this isomorphism is compatible with the Z_2-action and so induces an isomorphism of $M^+ \otimes 1$ with $M^- \otimes H$ over $B_{8q-r} \times P_{r-1}$. By the basic difference construction (see [2]) this defines a relative element

$$\mu_r \in KO(B_{8q} \times P_{r-1}, B_{8q-r} \times P_{r-1})$$

as required. Note that, from its construction, the image of α_r in the absolute group $KO(B_{8q} \times P_{r-1})$ is just $M^+ \otimes 1 - M^- \otimes H$.

Remarks. 1) A minor extension of the preceding construction leads to an element

$$\mu_{r,s} \in KO(B_{8q-s}/B_{8q-r-s} \divideontimes P_{r+s-1}/P_{s-1}).$$

This element is used for the other parts of Theorem (5.1) as well as for its generalizations. It is "natural" in both r and s.

2) For $r = 1$ we have

$$\mu_1 \in KO(B_{8q}, B_{8q-1}) = \widetilde{KO}(M \operatorname{Spin}(8q))$$

where $M \operatorname{Spin}(8q)$ is the universal Thom space over $B \operatorname{Spin}(8q)$. This element is the universal "Thom element" (see [2; § 12]).

Suppose now that Y is a Spin-manifold of dimension $8q$ with boundary ∂Y and that Y_0 is a closed subspace containing ∂Y. Assume that over Y_0 we have r linearly independent vector fields $v_1, \ldots, v_r$ of Y. Then the classifying map $Y \to B \operatorname{Spin} 8q$ becomes relativized to a map

$$f(v_1, \ldots, v_r) \colon (Y, Y_0) \to (B \operatorname{Spin}(8q), B \operatorname{Spin}(8q - r)).$$

Multiplying by P_{r-1} and putting $g = f \times \operatorname{Id}$ we then pull back the element μ_r to give an element

$$g^* \mu_r \in KO(Y \times P_{r-1}, Y_0 \times P_{r-1})$$

where $g = g(Y, Y_0; v_1, \ldots, v_r)$ depends on $v_1, \ldots, v_r$. In particular, taking Y to be the unit ball B^{8q} in R^{8q} and $Y_0 = \partial Y = S^{8q-1}$, we get an element of $KO^{-8q}(P_{r-1}) \cong KO(P_{r-1})$. This element depends on the vector fields $v_1, \ldots, v_r$ defined over S^{8q-1}, and our construction therefore defines a map

$$\alpha_r \colon \pi_{8q-1}(V_{8q,r}) \to KO(P_{r-1}).$$

A standard argument shows that this is a homomorphism. For $r = 1$ both groups are isomorphic to Z and, because μ_1 is the "Thom element", it follows that α_1 is an isomorphism. For $r = 2$ both groups are isomorphic to $Z \oplus Z_2$ and, because α_r is natural in r, the projections on Z coincide (via α_1). We shall see later that α_2 is actually an isomorphism and the projections on Z_2 also coincide.

Returning to our general Spin-manifold Y with $\partial Y \subset Y_0 \subset Y$ we recall that one can define a direct image homomorphism in KO-theory

$$KO(Y, \partial Y) \to KO(\text{point}) \cong Z$$

and more generally

$$KO(Y \times P, \partial Y \times P) \to KO(P)$$

for any compact auxiliary space P. This is done using the Thom isomorphism for Spin-bundles in KO-theory as explained in [3]. In particular, therefore, we have a homomorphism

$$KO(Y \times P_{r-1}, Y_0 \times P_{r-1}) \to KO(Y \times P_{r-1}, \partial Y \times P_{r-1}) \to KO(P_{r-1})$$

which we shall call the index because of its close connection with the index of elliptic operators. When $Y = B^{8q}$, $Y_0 = \partial Y = S^{8q-1}$ this index coincides with the periodicity isomorphism. Thus the element

(5.2) $\quad$ index $g(Y, Y_0; v_1, \ldots, v_r)^* \mu_r \in KO(P_{r-1})$

generalizes our homomorphism α_r.

Returning to our closed manifold X and the vector fields $v_1, \ldots, v_r$ with finite singularities at points $\{A\}$ we take $Y = X$ and $Y_0 = X - \bigcup_A B(A)$ where $B(A)$ is a small open ball around A. We then proceed to calculate the element (5.2) in two different ways. On the one hand, by excising the interior of Y_0, we see that it is equal to

$$\text{index } g \left(\bigcup_A \overline{B(A)}, \bigcup_A \partial B(A); v_1, \ldots, v_r \right)^* \mu_r$$
$$= \sum_A \text{index } g(\overline{B(A)}, \partial B(A); v_1, \ldots, v_r)^* \mu_r$$
(5.3) $\quad = \sum_A \alpha_r(a_A(v_1, \ldots, v_r)).$

Here we have used the additivity of the index for disjoint manifolds and the identification of α_r with index $g^* \mu$, already explained. On the other hand, since $\partial X = \emptyset$, we can also take $Y_0 = \emptyset$ and by the naturality of the index construction we have

$$\text{index } g(X, X - \bigcup_A (B(A); v_1, \ldots, v_r)^* \mu_r$$
$$= \text{index } g(X, \emptyset; v_1, \ldots, v_r)^* \mu_r$$

But now the index homomorphism involves only the absolute group $KO(X \times P_{r-1})$ and so we do not need $v_1, \ldots, v_r$. Moreover the element $g^* \mu_r$ is now just the element

$$\Delta^+(X) \otimes 1 - \Delta^-(X) \otimes H \in KO(X \times P_{r-1})$$

where $\Delta^{\pm}(X)$ denote the bundles associated to the principal Spin-bundle of X by the representations $\Delta^{\pm}$ of Spin$(8q)$. Thus

$$\text{index } g^* \mu_r = \text{index } (\Delta^+(X) \otimes 1) - \text{index } (\Delta^-(X) \otimes H)$$
$$= \text{index } \Delta^+(X) - (\text{index } \Delta^-(X)) \otimes H$$

(since index: $KO(X \times P_{r-1}) \to KO(P_{r-1})$ is a homomorphism of $KO(P_{r-1})$-modules). If we now apply the general index formula of [4] we find that*

$$\text{index } (\Delta^+(X) + \Delta^-(X)) = S(X)$$
$$\text{index } (\Delta^+(X) - \Delta^-(X)) = E(X)$$

and so

$$\text{index } \Delta^+(X) = \frac{E(X) + S(X)}{2}$$

$$\text{index } \Delta^-(X) = \frac{E(X) - S(X)}{2}.$$

Thus

$$\text{index } g^* \mu_r = \text{index } \Delta^+(X) - \text{index } \Delta^-(X)$$
$$- (\text{index } \Delta^-(X)) (H-1)$$
$$= E(X) + \frac{S(X) - E(X)}{2} (H-1).$$

Combining this with (5.3) we see that we have proved

(5.4) $$\sum_A \alpha_r(a_A(v_1, \ldots, v_r)) = E(X) + \frac{S(X) - E(X)}{2} (H-1).$$

This is a general formula for all r (for Spin-manifolds of dimension 8 q). We now put $r = 2$ to deduce the last part of Theorem (5.1). All that remains is to verify that α_2 is indeed an isomorphism as stated earlier. We shall do this by using (5.4).

We observe first that if X is a simply-connected 8 q-manifold then we can always find 2 vector fields v_1, v_2 with finite singularities and we can in fact assume there is just one singular point A. Apply this observation with $X = S^{8q}$ and we see that

$$\alpha_2(a_A(v_1, v_2)) = 2 - (H-1) \in KO(P_1).$$

---

* Our index: $KO(X) \to Z$ is the composition of the Thom isomorphism $\psi: KO(X) \cong KO(TX)$, complexification $KO \to K$ and the index homomorphism $K(TX) \to Z$ defined in [4]. One can then verify that $\psi(\Delta^+ + \Delta^-)$ is the symbol class of the operator D^+ of § 3 and similarly $\psi(\Delta^+ - \Delta^-)$ is the symbol class of the operator D. Alternatively, for the cohomologically minded, we can compute index $\Delta^\pm$ explicitly by the formulae in [3] and use the cohomological formulae for E, S.

But we already know (because of the relation between α_2 and α_1) that $1 + (H-1)$ is in the image of α_2. Hence α_2 is surjective and it follows easily that it is an isomorphism (Ker α_2 must be contained in the torsion subgroup and if not 0 then Im α_2 is cyclic). Finally we should check that the two "natural" decompositions of $\pi_{8k-1}(V_{8k,2})$ and $KO(P_1)$ correspond. For $KO(P_1)$ the natural decomposition is the one we have been implicitly using, namely $Z \oplus \widetilde{KO}(P_1)$, where Z corresponds to trivial bundles. For the group $\pi_{8q-1}(V_{8q,2})$ we use the fibration

$$S^{8q-2} \to V_{8q,2} \to S^{8q-1}$$

This is the 2-frame bundle of S^{8q-1} and a choice of non-zero vector field on S^{8q-1} gives a cross-section and hence splits the homotopy sequence

$$\begin{array}{ccccc} \pi_{8q-1}(S^{8q-2}) & \to & \pi_{8q-1}(V_{8q,2}) & \to & \pi_{8q-1}(S^{8q-1}) \\ \| & & & & \| \\ Z_2 & & & & Z \end{array}$$

We get a standard vector field on S^{8q-1} by regarding it as the unit sphere in C^{4q} and using multiplication by i. If we let b be the element of $\pi_{8q-1}(V_{8q,2})$ given by this cross-section we have to verify that

$$\alpha_2(b) = 1 \in KO(P_1)$$

(and not $1 + (H-1)$, which is the only possible alternative). The easiest way to verify this is to use (5.4) for the complex projective space $P_{4q}(C)$. Strictly speaking we have not proved (5.4) in this case because $P_{4q}(C)$ is not a Spin-manifold, but we will ignore this point (which will in any case be taken up more systematically elsewhere). So let v_1 be a holomorphic vector field on $P_{4q}(C)$ with just $4q+1$ zeros (given by a general projective linear transformation) and put $v_2 = iv_1$. Then for each singularity A we have $a_A(v_1, v_2) = b$ and so

(5.5) $$\sum_A \alpha_2 a_A(v_1, v_2) = (4q+1)\alpha_2(b).$$

But for $X = P_{4q}(C)$ we have

$$E(X) = 4q+1, \ S(X) = 1$$

and hence from (5.4) and (5.5) we deduce $\alpha_2(b) = 1$ as stated.

References

Topology of Elliptic Operators

[1] M. F. Atiyah, Amer. Math. Symposium on Global Analysis (Berkeley, 1968).
[2] M. F. Atiyah, R. Bott and A. A. Shapiro, Clifford modules, Topology, 3 (Suppl. 1) (1964), 3–38.
[3] M. F. Atiyah and F. Hirzebruch, Riemann–Roch theorems for differentiable manifolds, Bull. Amer. Math. Soc. 65 (1959), 276–281.
[4] M. F. Atiyah and I. M. Singer, The index of elliptic operators: I, Ann. of Math. 87 (1968), 484–530.
[5] –, The index of elliptic operators: III, Ann. of Math. 87 (1968), 546–604.
[6] –, Index theory for skew-adjoint Fredholm operators, Publ. Math. I.H.E.S. (1970) (to appear).
[7] D. Frank (to appear).
[8] G. Lusztig, J. Milnor and F. Peterson, Semi-characteristics and cobordism, Topology, 8 (1969), 357–359.
[9] K. H. Mayer, Elliptische Differentialoperatoren und Ganzzahligkeitssätze für charakteristische Zahlen, Topology, 4 (1965), 295–313.
[10] E. Thomas, Vector fields on manifolds, Bull. Amer. Math. Soc. 75 (1969), 643–683.

Zusammenfassung

In der Arbeit wird insbesondere die Frage nach der Existenz von r linearunabhängigen Vektorfeldern auf Mannigfaltigkeiten untersucht. Es werden nicht die üblichen topologischen Methoden angewandt, die Untersuchung basiert vielmehr auf den neuen Ergebnissen von Atiyah und Singer über elliptische Operatoren. Ausgangspunkt ist die Tatsache, daß klassische topologische Invarianten einer Mannigfaltigkeit, zum Beispiel die Eulersche Charakteristik und die Signatur, Indizes elliptischer Operatoren sind und die Existenz einer gewissen Anzahl von Vektorfeldern Symmetrieeigenschaften dieser Operatoren und damit auch besondere Aussagen über ihre Indizes impliziert. Auf diese Weise ergeben sich notwendige Bedingungen für die Existenz von Vektorfeldern.

Résumé

Dans ce travail est notamment examinée la question de l'existence de r champs de vecteurs linéairement indépendants sur une variété. On n'utilise pas les méthodes topologiques usuelles mais plutôt les nouveaux résultats d'Atiyah et Singer sur les opérateurs elliptiques. Le point de départ est le fait que des invariants topologiques classiques d'une variété, par exemple la caractéristique d'Euler-Poincaré et la signature, sont les indices d'opérateurs elliptiques, et que l'existence d'un certain nombre de champs de vecteurs implique des propriétés de symétrie pour ces opérateurs et, par là, certaines assertions concernant leurs indices. De cette façon sont obtenues des conditions nécessaires d'existence de champs de vecteurs.

VERÖFFENTLICHUNGEN DER ARBEITSGEMEINSCHAFT FÜR FORSCHUNG DES LANDES NORDRHEIN-WESTFALEN

Neuerscheinungen 1967 bis 1970

AGF-N Heft Nr.		NATUR-, INGENIEUR- UND GESELLSCHAFTSWISSENSCHAFTEN
165	*Reimar Lüst, Garching*	Weltraumforschung in der Bundesrepublik und Europa
	Karl-Otto Kiepenheuer, Freiburg i. Br.	Sonnenforschung
166	*Amos de-Shalit, Rehovoth (Israel)*	Die naturwissenschaftliche Forschung in kleinen Ländern. Das Beispiel Israels
167	*Ernst Derra, Düsseldorf*	Die Herz- und Herzgefäßchirurgie im derzeitigen Stadium
	Franz Grosse-Brockhoff, Düsseldorf	Elektrotherapie von Herzerkrankungen
168	*Hans Hermes, Freiburg i. Br.*	Die Rolle der Logik beim Aufbau naturwissenschaftlicher Theorien
169	*Friedrich Mölbert, Hannover*	Wechselbeziehungen zwischen Biologie und Technik
	Dietrich Schneider, Seewiesen üb. Starnberg	Die Arbeitsweise tierischer Sinnesorgane im Vergleich zu technischen Meßgeräten
170	*John Flavell Coales, Cambridge (England)*	Automation und Computer in der Industrie
	Ludwig Pack, Münster	Raumzuordnung und Raumform von Büro- und Fabrikgebäuden
171	*Wilhelm Menke, Köln*	Die Struktur der Chloroplasten
	Achim Trebst, Göttingen	Zum Mechanismus der Photosynthese
172	*Heinrich Heesch, Hannover*	Reguläres Parkettierungsproblem
173	*Wilhelm Becker, Basel*	Das Milchstraßensystem als spiralförmiges Sternsystem
	Hans Haffner, Hamburg	Sternhaufen und Sternentwicklung
174	*Karl-Heinrich Bauer, Heidelberg*	Vom Krebsproblem – heute und morgen
	Richard Haas, Freiburg i. Br.	Virus und Krebs
175	*Karlheinz Althoff, Bonn*	Von 500 MeV zu 2500 MeV – Entwicklung der Hochenergiephysik in Bonn
	Theo Mayer-Kuckuk, Bonn	Kernstrukturuntersuchungen mit modernen Beschleunigern
176	*Michael Grewing, Jörg Pfleiderer und Wolfgang Priester, alle Bonn*	Nichtthermische kosmische Strahlungsquellen
177	*Otto Hachenberg, Bonn*	Betrachtungen zum Bau großer Radioteleskope
178	*Uichi Hashimoto, Tokyo*	Die Eisen- und Stahlindustrie in Japan
179	*Paul Klein, Mainz*	Humorale Mechanismen der immunbiologischen Abwehrleistungen
	Herbert Fischer, Freiburg i. Br.	Zelluläre Aspekte der Immunität
	Ernst Friedrich Pfeiffer, Ulm	Immunologische Aspekte der modernen Endokrinologie
180	*Benno Hess, Dortmund*	Probleme der Regulation zellulärer Prozesse
	Norbert Weissenfels, Bonn	Die Gewebezüchtung im Dienste der experimentellen Zellforschung
181	*Josef Meixner, Aachen*	Beziehungen zwischen Netzwerktheorie und Thermodynamik
	Friedrich Schlögl, Aachen	Informationstheorie und Thermodynamik irreversibler Prozesse

182	Wilhelm Dettmering, Aachen	Entwicklungslinien der luftansaugenden Strahltriebwerke
183	Hermann Merxmüller, München	Moderne Probleme der Pflanzensystematik
	Hans Mohr, Freiburg i. Br.	Die Streuung der Entwicklung durch das Phytochromsystem
184	Frederik van der Blij, Utrecht	Zahlentheorie in Vergangenheit und Zukunft
	Georges Papy, Brüssel	Der Einfluß der mathematischen Forschung auf den Schulunterricht
185	Rudolf Schulten, Jülich	Zukünftige Anwendung der nuklearen Wärme
	Günther Dibelius, Aachen	
	Werner Wenzel, Aachen	
186	Friedrich Becker, München	Ausblick in das Weltall
187	Kuno Radius, Konstanz	Probleme der Entwicklung von Großrechenanlagen
	Hans Kaufmann, München	Speicher- und Schaltkreis-Technik von Daten-Verarbeitungs-Anlagen
	Hans Jörg Tafel, Aachen	Strömungsmechanische Nachrichtenübertragung und -verarbeitung (Fluidik)
188	Erwin Bodenstedt, Bonn	Beobachtung der Resonanz zwischen elektrischer und magnetischer Hyperfeinstruktur-Wechselwirkung
	Siegfried Penselin, Bonn	Probleme der Zeitmessung
189	August Wilhelm Quick, Aachen	Die dritte Stufe der europäischen Trägerrakete unter besonderer Berücksichtigung der Prüfung durch Höhensimulationsanlagen
	Philipp Hartl, Oberpfaffenhofen	Der deutsche Forschungssatellit und der deutsch-französische Nachrichtensatellit
	Werner Fogy, Oberpfaffenhofen	Das deutsche Bodenstationssystem für den Funkverkehr mit Satelliten
190	Sir Denning Pearson, Derby	Probleme der Unternehmensführung in der internationalen Flugtriebwerksindustrie
	Lord Jackson of Burnley, London	Die Abwanderung von qualifizierten Fachkräften
191	Hans Ebner, Aachen	Konstruktive Probleme der Ozeanographischen Forschung
192	Harald Schäfer, Münster	Verbindungen der schweren Übergangsmetalle mit Metall—Metall-Bindungen
	Hans Musso, Bochum und Marburg	Über die Struktur organischer Metallkomplexe
193	Friedrich Seidel, Marburg a. d. Lahn	Entwicklungspotenzen des frühen Säugetierkeimes
	Robert Domenjoz, Bonn	Die entzündliche Reaktion und die antiphlogistischen Heilmittel
194	Eugen Flegler, Aachen	Probleme des elektrischen Durchschlags
195	Franz Lotze, Münster	Die Salz-Lagerstätten in Zeit und Raum Ein Beitrag zum Klima der Vorzeit
196	Johann Schwartzkopff, Bochum	Die Verarbeitung von akustischen Nachrichten im Gehirn von Tieren verschiedener Organisationshöhen
	Werner Kloft, Bonn	Radioaktive Isotope und ionisierende Strahlung bei der Erforschung und Bekämpfung von Insekten
197	Werner Heinrich Hauss, Münster	Über Entstehung und Verhütung der Arteriosklerose
	Hans-Werner Schlipköter, Düsseldorf	Ätiologie und Pathogenese der Silikose sowie ihre kausale Beeinflussung
198	Louis Néel, Grenoble	Elementarbezirke und Wände in einem ferromagnetischen Kristall
199	J. Herbert Hollomon, Norman, Okl.	Systems Management
	Stewart Blake, Menlo Park, Kalifornien	
	Emanuel R. Piore, New York	
	Wilhelm Krelle, Bonn	
	David B. Hertz, New York	
200	Michael F. Atiyah	Vector Fields on Manifolds
201	Jan Tinbergen, Rotterdam	Optimale Produktionsstruktur und Forschungsrichtung
	Hans A. Havemann, Aachen	Neue Aspekte der Entwicklungsländerforschung
202	Pe'er Mittelstaedt, Köln	Lorentzinvariante Gravitationstheorie

AGF-WA WISSENSCHAFTLICHE ABHANDLUNGEN
Band Nr.

1	Wolfgang Priester, Hans-Gerhard Bennewitz und Peter Lengrüßer, Bonn	Radiobeobachtungen des ersten künstlichen Erdsatelliten
2	Joh. Leo Weisgerber, Bonn	Verschiebungen in der sprachlichen Einschätzung von Menschen und Sachen
3	Erich Meuthen, Marburg	Die letzten Jahre des Nikolaus von Kues
4	Hans-Georg Kirchhoff, Rommerskirchen	Die staatliche Sozialpolitik im Ruhrbergbau 1871–1914
5	Günther Jachmann, Köln	Der homerische Schiffskatalog und die Ilias
6	Peter Hartmann, Münster	Das Wort als Name (Struktur, Konstitution und Leistung der benennenden Bestimmung)
7	Anton Moortgat, Berlin	Archäologische Forschungen der Max-Freiherr-von-Oppenheim-Stiftung im nördlichen Mesopotamien 1956
8	Wolfgang Priester und Gerhard Hergenhahn, Bonn	Bahnbestimmung von Erdsatelliten aus Doppler-Effekt-Messungen
9	Harry Westermann, Münster	Welche gesetzlichen Maßnahmen zur Luftreinhaltung und zur Verbesserung des Nachbarrechts sind erforderlich?
10	Hermann Conrad und Gerd Kleinheyer, Bonn	Vorträge über Recht und Staat von Carl Gottlieb Svarez (1746–1798)
11	Georg Schreiber†, Münster	Die Wochentage im Erlebnis der Ostkirche und des christlichen Abendlandes
12	Günther Bandmann, Bonn	Melancholie und Musik. Ikonographische Studien
13	Wilhelm Goerdt, Münster	Fragen der Philosophie. Ein Materialbeitrag zur Erforschung der Sowjetphilosophie im Spiegel der Zeitschrift „Voprosy Filosofii" 1947–1956
14	Anton Moortgat, Berlin	Tell Chuéra in Nordost-Syrien. Vorläufiger Bericht über die Grabung 1958
15	Gerd Dicke, Krefeld	Der Identitätsgedanke bei Feuerbach und Marx
16a	Helmut Gipper, Bonn, und Hans Schwarz, Münster	Bibliographisches Handbuch zur Sprachinhaltsforschung, Teil I. Schrifttum zur Sprachinhaltsforschung in alphabetischer Folge nach Verfassern – mit Besprechungen und Inhaltshinweisen (Erscheint in Lieferungen: bisher Bd. I, Lfg. 1–7; Lfg. 8–10)
17	Thea Buyken, Bonn	Das römische Recht in den Constitutionen von Melfi
18	Lee E. Farr, Brookhaven, Hugo Wilhelm Knipping, Köln, und William H. Lewis, New York	Nuklearmedizin in der Klinik. Symposion in Köln und Jülich unter besonderer Berücksichtigung der Krebs- und Kreislaufkrankheiten
19	Hans Schwippert, Düsseldorf, Volker Aschoff, Aachen, u. a.	Das Karl-Arnold-Haus. Haus der Wissenschaften der Arbeitsgemeinschaft für Forschung des Landes Nordrhein-Westfalen in Düsseldorf. Planungs- und Bauberichte (Herausgegeben von Leo Brandt, Düsseldorf)
20	Theodor Schieder, Köln	Das deutsche Kaiserreich von 1871 als Nationalstaat
21	Georg Schreiber†, Münster	Der Bergbau in Geschichte, Ethos und Sakralkultur
22	Max Braubach, Bonn	Die Geheimdiplomatie des Prinzen Eugen von Savoyen
23	Walter F. Schirmer, Bonn, und Ulrich Broich, Göttingen	Studien zum literarischen Patronat im England des 12. Jahrhunderts
24	Anton Moortgat, Berlin	Tell Chuéra in Nordost-Syrien. Vorläufiger Bericht über die dritte Grabungskampagne 1960
25	Margarete Newels, Bonn	Poetica de Aristoteles traducida de latin. Ilustrada y comentada por Juan Pablo Martir Rizo (erste kritische Ausgabe des spanischen Textes)
26	Vilho Niitemaa, Turku, Pentti Renvall, Helsinki, Erich Kunze, Helsinki, und Oscar Nikula, Åbo	Finnland – gestern und heute
27	Ahasver von Brandt, Heidelberg, Paul Johansen, Hamburg, Hans van Werveke, Gent, Kjell Kumlien, Stockholm, Hermann Kellenbenz, Köln	Die Deutsche Hanse als Mittler zwischen Ost und West

28	*Hermann Conrad, Gerd Kleinheyer, Thea Buyken und Martin Herold, Bonn*	Recht und Verfassung des Reiches in der Zeit Maria Theresias. Die Vorträge zum Unterricht des Erzherzogs Joseph im Natur- und Völkerrecht sowie im Deutschen Staats- und Lehnrecht
29	*Erich Dinkler, Heidelberg*	Das Apsismosaik von S. Apollinare in Classe
30	*Walther Hubatsch, Bonn, Bernhard Stasiewski, Bonn, Reinhard Wittram, Göttingen, Ludwig Petry, Mainz, und Erich Keyser, Marburg (Lahn)*	Deutsche Universitäten und Hochschulen im Osten
31	*Anton Moortgat, Berlin*	Tell Chuēra in Nordost-Syrien. Bericht über die vierte Grabungskampagne 1963
32	*Albrecht Dihle, Köln*	Umstrittene Daten. Untersuchungen zum Auftreten der Griechen am Roten Meer
33	*Heinrich Behnke und Klaus Kopfermann (Hrsgb.), Münster*	Festschrift zur Gedächtnisfeier für Karl Weierstraß 1815–1965
34	*Joh. Leo Weisgerber, Bonn*	Die Namen der Ubier
35	*Otto Sandrock, Bonn*	Zur ergänzenden Vertragsauslegung im materiellen und internationalen Schuldvertragsrecht. Methodologische Untersuchungen zur Rechtsquellenlehre im Schuldvertragsrecht
36	*Iselin Gundermann, Bonn*	Untersuchungen zum Gebetbüchlein der Herzogin Dorothea von Preußen
37	*Ulrich Eisenhardt, Bonn*	Die weltliche Gerichtsbarkeit der Offizialate in Köln, Bonn und Werl im 18. Jahrhundert
38	*Max Braubach, Bonn*	Bonner Professoren und Studenten in den Revolutionsjahren 1848/49
39	*Henning Bock (Bearb.), Berlin*	Adolf von Hildebrand Gesammelte Schriften zur Kunst
40	*Geo Widengren, Uppsala*	Der Feudalismus im alten Iran

Sonderreihe
PAPYROLOGICA COLONIENSIA

Vol. I
Aloys Kehl, Köln   Der Psalmenkommentar von Tura, Quaternio IX
                                  (Pap. Colon. Theol. 1)

Vol. II
Erich Lüddeckens, Würzburg
P. Angelicus Kropp O. P. †, Klausen   Demotische und
Alfred Hermann und Manfred Weber, Köln Koptische Texte

Vol. III
Stephanie West, Oxford   The Ptolemaic Papyri of Homer

Vol. IV
Ursula Hagedorn und
Dieter Hagedorn, Köln,   Das Archiv des Petaus (P. Petaus)
Louise C. Youtie und
Herbert C. Youtie, Ann Arbor
(Hrsgb.)

SONDERVERÖFFENTLICHUNGEN

Herausgeber: Der Ministerpräsident
des Landes Nordrhein-Westfalen   Jahrbuch 1963, 1964, 1965, 1966, 1967, 1968 und 1969
– Landesamt für Forschung –   des Landesamtes für Forschung

Verzeichnisse sämtlicher Veröffentlichungen der Arbeitsgemeinschaft
für Forschung des Landes Nordrhein-Westfalen können beim
Westdeutschen Verlag GmbH, 567 Opladen, Ophovener Str. 1–3, angefordert werden.

MIX
Papier aus verantwortungsvollen Quellen
Paper from responsible sources
FSC® C105338

If you have any concerns about our products,
you can contact us on
ProductSafety@springernature.com

In case Publisher is established outside the EU,
the EU authorized representative is:
**Springer Nature Customer Service Center GmbH
Europaplatz 3, 69115 Heidelberg, Germany**

Printed by Libri Plureos GmbH
in Hamburg, Germany